Kate Peridot

Samara Hardy

Animal Swim School

BLOOMSBURY
CHILDREN'S BOOKS

LONDON OXFORD NEW YORK NEW DELHI SYDNEY

Did you know that some animals are **natural swimmers**? If a **cat**, a **dog**, a **duck** or a **frog** fall into a pool, even if they don't like water, they automatically know how to paddle. But humans don't have this same skill – we need to **learn** to swim, and we do this by copying others.

So, who first showed humans how to **float** and **move** about in the water?

Animals of course!

Welcome to the Animal Swim School!

These animal friends are here to teach you everything you need to know about being **safe** in the water, feeling **brave** and learning your first swim strokes.

What are you waiting for? Let's meet the **animal swimming experts**!

Remember to always be with a grown-up when you're near water.

Wallow like a hippo

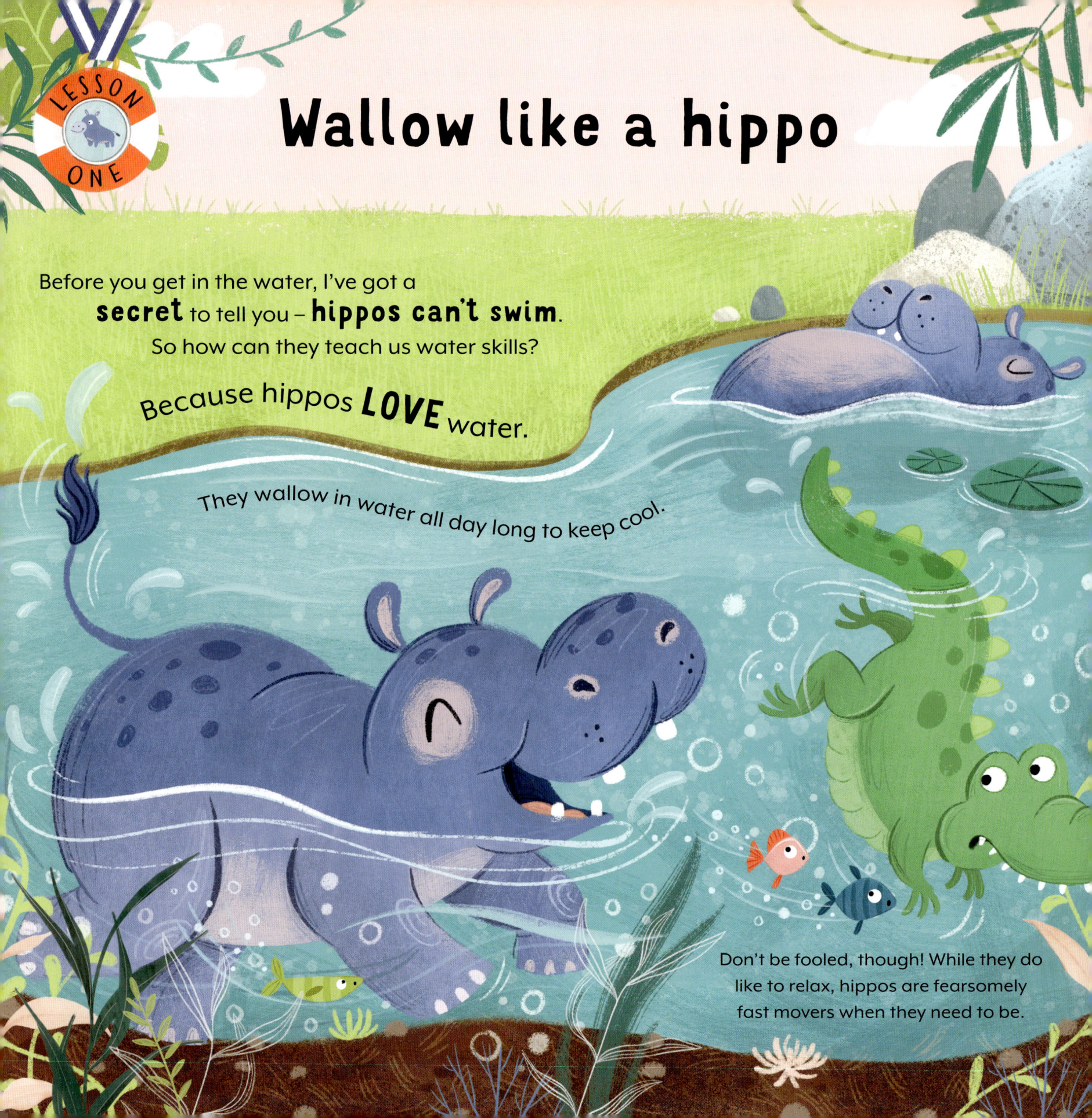

Before you get in the water, I've got a **secret** to tell you – **hippos can't swim**. So how can they teach us water skills?

Because hippos **LOVE** water.

They wallow in water all day long to keep cool.

Don't be fooled, though! While they do like to relax, hippos are fearsomely fast movers when they need to be.

1 Step into shallow water with a **big hippo SPLASH**.

2 Sink so only your **nose**, **eyes** and **ears** are above the surface.

3 Now **stride** across the pool!

If water goes up your nose,

SNORT!

Equipment like armbands, floats, woggles and goggles are very helpful when you're learning to swim.

Hold on like a crab

When the water gets too **splishy splashy**, crabs know it's time to **scuttle** onto a rock and take a break. They have strong claws that are very handy for **holding on**.

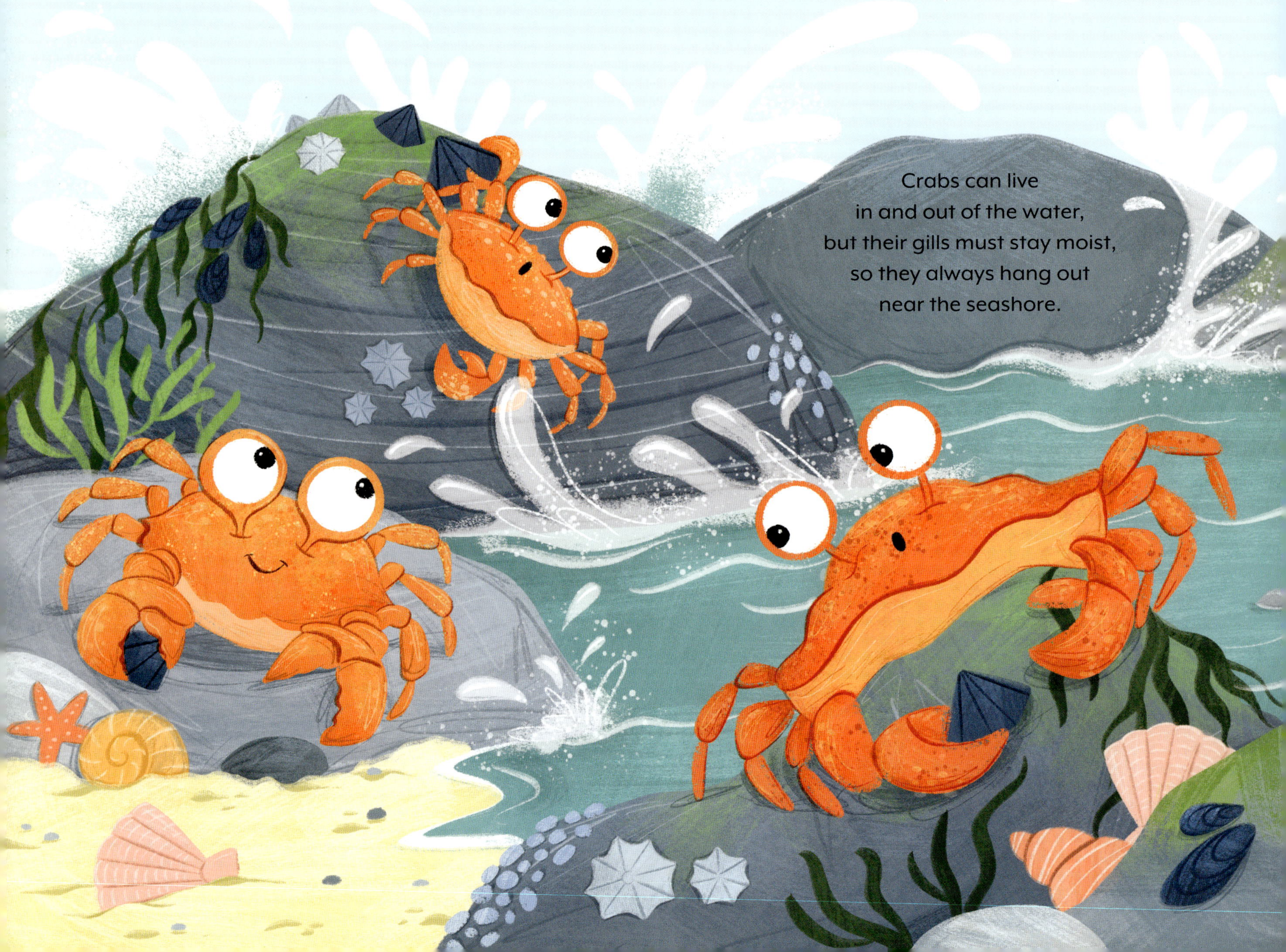

Crabs can live in and out of the water, but their gills must stay moist, so they always hang out near the seashore.

Are your **pincer fingers** ready?

SNAP, SNAP!

Hold on to the edge and **scuttle sideways** around the pool.
Can you spot a safe place to climb out?

Float on your back like a sea otter

Sea otters spend nearly all their time in the water. When they want to take a break from swimming, they **float** on their backs.

Otters are very buoyant, which means floaty. They even eat and sleep while floating. Sometimes families hold paws so they don't drift apart.

1 **Lie back** in the water and relax, letting your arms float slightly out from your sides. Look up and **push** your tummy up.

2 Relax and **breathe normally**. The more you relax the easier you'll float!

Blow bubbles like a sea lion

Sea lions are the acrobats of the ocean. They blow bubbles for fun, but this also helps them to **dive**.

When sea lions blow bubbles, they release air, which makes it easier to sink. It also stops water from going up their noses as they twist and roll.

1 In shallow water, take a **deep breath** and duck your head just below the surface. Blow out through your nose.

Blub,
blub,
blub.

Feel your body
slowly sink.

2 Then launch upwards suddenly to take a breath. **SURPRISE!**

Float on your front like a platypus

When platypuses aren't snoozing in their burrows, they **float** on the surface of the river with their webbed **paws outstretched**. They look down to spy creatures in the deep who might make a tasty snack.

Yum, yum, yum!

Platypuses swim with their eyes closed when they are hunting for food. They love to eat crunchy shrimps.

1 Take a breath and put your face in the water. **Stretch** your arms out a little and let your legs **rise up** behind you.

2 While you're floating, **open your eyes** and look down. What can you see?

Paddle like a dog

Most dogs just can't help it, every time they see water, they get so **excited** they **have** to jump in!

SPLASH!

Doggy-paddle isn't the fastest swim stroke, but you'll always make it across the pool with your nose above the water.

1 Make a scoop with your hand and paddle the water, one hand after the other – **forward**, **down** and **under**.

2 Kick your legs **up** and **down** with pointed toes.

This is called a **flutter kick**.

3 Now **jump** out of the water and **spray** everyone by shaking out your wet hair.

Woof!

Shoot off like a squid

Squid love to drift with the current, but when they want to zoom ... **WHOOSH** ... off they go!

They gather their tentacles into a point to create a smooth shape that **glides** easily through the water.

Squid can squirt a jet of water out of their bodies to shoot off quickly.

1 **Hold** on to the side of the pool with one hand.

2 **Point** your other arm in front of you.

3 Place both feet on the wall and **push off**.

4 Swing your back arm over to join the front, making an **arrow shape** with your hands.

How far can you go?

Keep your body straight.

Point your toes!

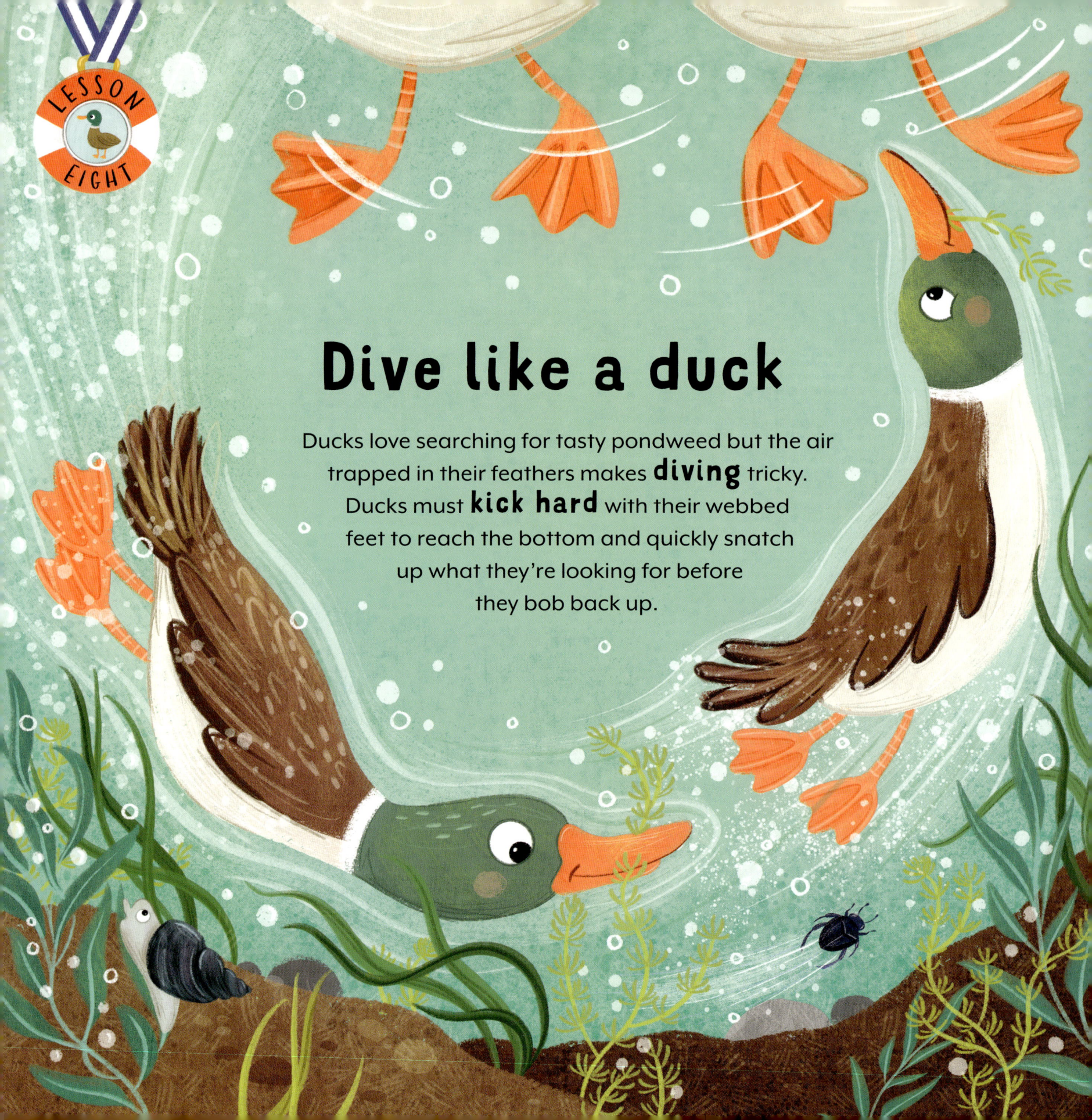

Dive like a duck

Ducks love searching for tasty pondweed but the air trapped in their feathers makes **diving** tricky. Ducks must **kick hard** with their webbed feet to reach the bottom and quickly snatch up what they're looking for before they bob back up.

1 Take a breath, **bend** at the waist and **point** your arms down.

4 To come back up, **bounce** off the bottom with your feet and kick.

Dive to collect pool toys for fun!

2 Blow bubbles slowly out of your nose and **push** the water away with your arms – like froggy arms.

3 Flutter **kick** your legs like doggy kicks.

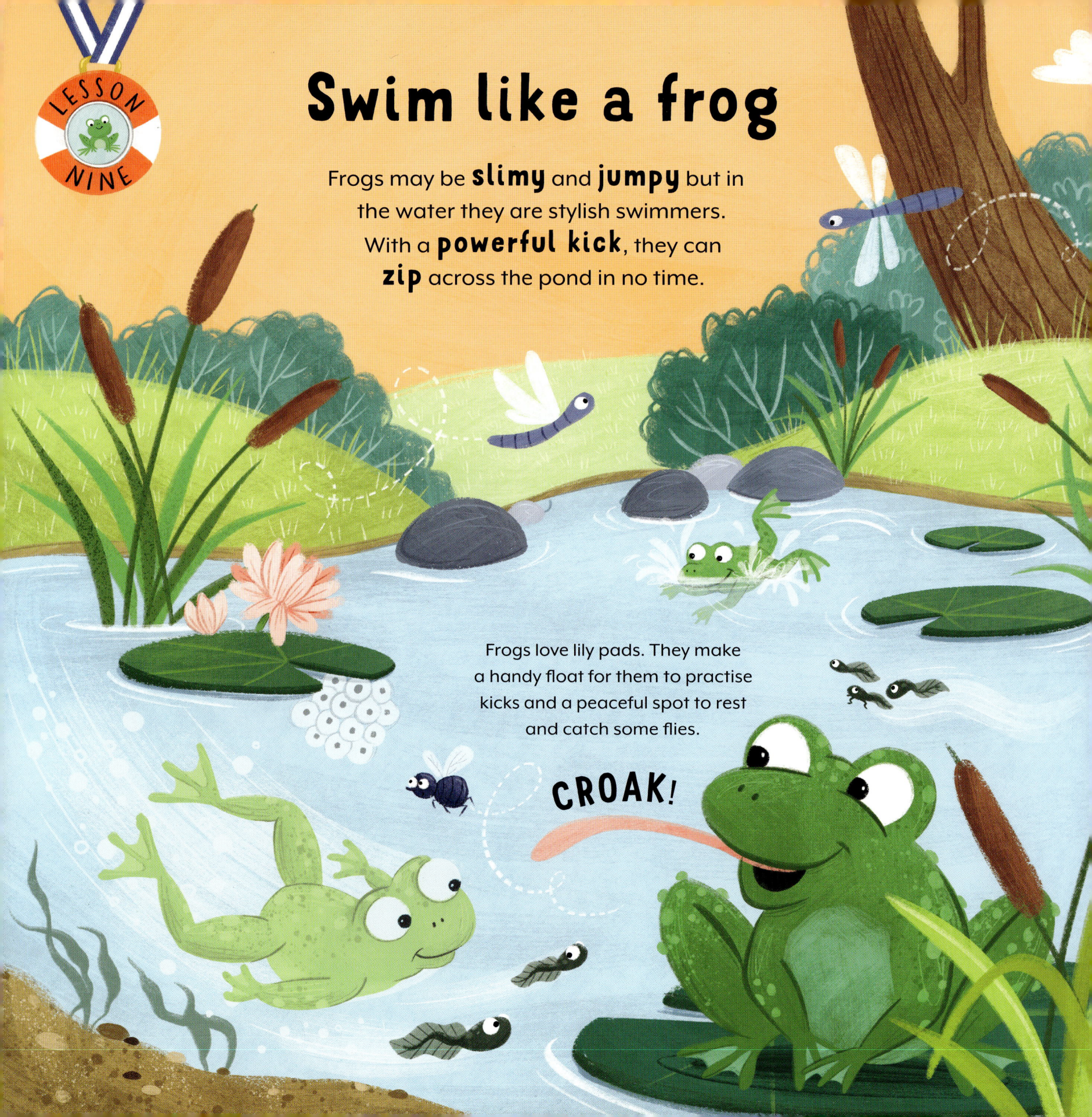

Swim like a frog

Frogs may be **slimy** and **jumpy** but in the water they are stylish swimmers. With a **powerful kick**, they can **zip** across the pond in no time.

Frogs love lily pads. They make a handy float for them to practise kicks and a peaceful spot to rest and catch some flies.

CROAK!

1 **Stretch** your arms in front with palms facing outward, fingers together.

2 Put your face in the water and **push** the water sideways and back with your arms.

3 As your arms come level with your shoulders, **lift** your head and take a breath.

4 **Pull** your knees up beneath you.

5 **Kick** out to the side and back, then bring your feet together. Then bring your hands together beneath you and repeat.

This stroke is called breaststroke.

Tread water like a seahorse

Seahorses are tiny creatures that swim **upright**. When ocean currents push them around, they **flutter** their fins super-fast to stay in one spot, and when they're tired, they loop their tails around a piece of seaweed to rest.

1 In shoulder-high water, **raise** and **bend** your elbows, then **push your palms** out to the side and in again in a paddling movement.

This is called sculling.

2 Lift your knees up and **frog kick** downwards one leg at a time in a circular motion.

Leaning slightly forward will help you to float.

Snorkel like an elephant

You're probably thinking that elephants are too big and heavy to float. Well you'd be wrong! Elephants can **float** with their heads just beneath the surface. Fortunately, they have their own **inbuilt snorkel** – their trunk!

Once back in the shallows, everyone in the herd enjoys a good hosing down.

SPLISH, SPLASH!

1 Put on your mask and snorkel. **Duck** underwater to check your mask doesn't leak and the snorkel **points up** in the air.

2 Swim **looking down** and breathe normally through your mouth. Your breath travels through the snorkel. **What can you see?**

3 When you come up, **blow** hard through your snorkel to **clear** any water.

Race like a dolphin

Welcome to the dolphin gang – known as the pod. Dolphins are **playful** and **fast**.
With strong beats of their tail, they can leap above the waves.

Sleek dolphins are built for swimming!

Pods of dolphins race against each other, whales and even boats!

1 Keep your **legs** and **feet together** and imagine you have a **dolphin tail**.

2 **Roll** your body from your **chest** to your **hips** and **flick** your **feet**. Put your head up to take a breath any time.

3 Then try this on your **back**. Which is easier?

Practise all the animal swimming skills!

Water confidence and safety skills

Wallow like a hippo

Hold on like a crab

Float on your back like a sea otter

Blow bubbles like a sea lion

First swim skills

Float on your front like a platypus

Paddle like a dog

Shoot off like a squid

Dive like a duck

More advanced swim skills

Swim like a frog

Tread water like a seahorse

Snorkel like an elephant

Race like a dolphin

CONGRATULATIONS, you've tried lots of these different skills and you're on your way to becoming an amazing swimmer!

Now you're ready to swim like a human!

People have taken inspiration from animals and invented their own swim strokes.

Front crawl (or freestyle)
Swimmers swim on their fronts. They flutter kick with their legs and move their arms like a wheel, forward, under and round, one arm and then the other. They turn their body and head when they lift an arm to catch a breath.

Backstroke
This is the same technique as front crawl, but swimmers are on their backs.

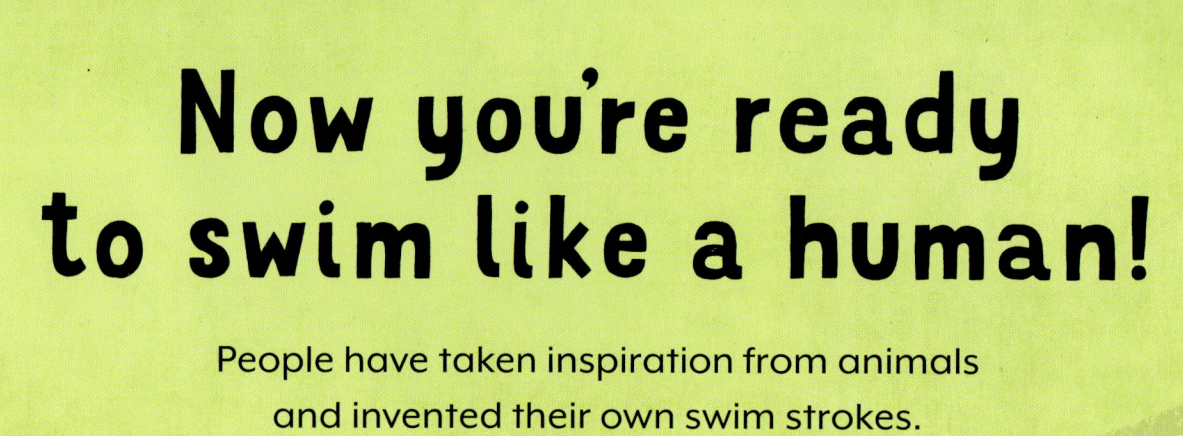

Butterfly stroke
Swimmers move their arms like the beat of a butterfly's wings and move their body and legs like a dolphin.

Breaststroke
This is the proper name for swimming like a frog.

People swim these strokes in competitions. Front crawl is the fastest human swim stroke.

Give these human strokes a try. Which is your favourite?

For Jessica and Jack, who grew up to swim like fish. — K.P.
For Jean and Joan — S.H.

BLOOMSBURY CHILDREN'S BOOKS
Bloomsbury Publishing Plc
50 Bedford Square, London, WC1B 3DP, UK
29 Earlsfort Terrace, Dublin 2, Ireland

BLOOMSBURY, BLOOMSBURY CHILDREN'S BOOKS and the Diana logo
are trademarks of Bloomsbury Publishing Plc

First published in Great Britain 2024 by Bloomsbury Publishing Plc

A catalogue record for this book is available from the British Library

ISBN: PB: 978-1-5266-5699-5; eBook: 978-1-5266-7283-4
2 4 6 8 10 9 7 5 3 1

Printed in China by RR Donnelley, Dongguan City, Guangdong

MIX
Paper | Supporting responsible forestry
FSC
www.fsc.org
FSC® C144853

To find out more about our authors and books visit
www.bloomsbury.com and sign up for our newsletters